Mathematics for Classical Information Retrieval

Mathematics

for

Classical Information Retrieval:

Roots and Applications

Dariush Alimohammadi

Department of Library and Information Studies
Tarbiat Moallem University, Tehran, Iran

Edited by Dr. Mary K. Bolin

Professor and Chair of Technical Services
University Libraries, University of Nebraska–Lincoln,
Lincoln, Nebraska, USA

Zea E-Books
2010

ISBN 978-1-60962-003-5   (electronic)
ISBN 978-1-60962-004-2   (paperback)

Zea E-Books are published by the University of Nebraska–Lincoln Libraries,
Lincoln, Nebraska.

To my dear wife,

Mahshid Sajjadi,

for her infinite help and patience

— Dariush Alimohammadi

Contents

Table of Notations

In the table of notations, the most popular mathematical symbols used through the text have been defined according to their presence at the book.

Symbol	Name	Meaning
=	equality	$x = y$ means x and y represent the same thing or value.
{ , }	set brackets	{a,b,c} means the set consisting of a, b, and c.
:	extends; over	K : F means the field K extends the field F.
<	is less than	$x < y$ means x is less than y.
≤	is less than or equal to	$x \leq y$ means x is less than or equal to y.
>	is greater than	$x > y$ means x is greater than y.
≥	is greater than or equal to	$x \geq y$ means x is greater than or equal to y.
( )	parentheses	Perform the operations inside the parentheses first.
∈	is an element of	$a \in S$ means a is an element of the set S.
∉	is not an element of	$a \notin S$ means a is not an element of S.
∅	the empty set	∅ means the set with no elements.
∪	the union of	A ∪ B means the set of those elements which are either in A, or in B, or in both.
∩	intersected with	A ∩ B means the set that contains all those elements that A and B have in common.
−	negative; minus; the opposite of	−3 means the negative of the number 3.
×	the Cartesian product of … and …; the direct product of … and …	X × Y means the set of all ordered pairs with the first element of each pair selected from X and the second element selected from Y.
∧	and; min; meet	The statement A ∧ B is true if A and B are both true; else it is false.
∨	or; max; join	The statement A ∨ B is true if A or B (or both) are true; if both are false, the statement is false.
⇒	implies; if … then	A ⇒ B means if A is true then B is also true; if A is false then nothing is said about B.
⇔	if and only if	A ⇔ B means A is true if B is true and A is false if B is false.

Symbol	Name	Meaning
|	conditional probability	$P(A \mid B)$ means the probability of the event A occurring given that B occurs.
$\sum$	sum over ... from ... to ... of	$\sum_{k=1}^{n} a_k$ means $a_1 + a_2 + \ldots + a_n$.
[ ]	1 if true, 0 otherwise	[S] maps a true statement S to 1 and a false statement S to 0.
$\mathbb{R}$	R; the (set of) real numbers; the reals	$\mathbb{R}$ means the set of real numbers.
$\int$	integral from ... to ... of ... with respect to	$\int_a^b f(x)\,dx$ means the signed area between the x-axis and the graph of the function f between $x = a$ and $x = b$.
$\sqrt{\ }$	the (principal) square root of	$\sqrt{x}$ means the positive number whose square is x.
∞	Infinity	∞ is an element of the extended number line that is greater than all real numbers; it often occurs in limits.
$\subset$	is a subset of	(proper subset) $A \subset B$ means $A \subseteq B$ but $A \neq B$.
$\perp$	orthogonal/perpendicular complement of; perp	$W^\perp$ means the orthogonal complement of W (where W is a subspace of the inner product space V), the set of all vectors in V orthogonal to every vector in W.
δ	Dirac delta of	$\delta(x) = \begin{cases} \infty, & x = 0 \\ 0, & x \neq 0 \end{cases}$
$\forall$	for all; for any; for each	$\forall\, x : P(x)$ means $P(x)$ is true for all x.
$\mathbb{N}$	the (set of) natural numbers	$\mathbb{N}$ means either $\{\,0, 1, 2, 3, \ldots\}$.
$\mathbb{Q}$	the (set of) rational numbers; the rationals	$\mathbb{Q}$ means $\{p/q : p \in \mathbb{Z}, q \in \mathbb{N}\}$.
Δ	delta; change in	Δx means a (non-infinitesimal) change in x.
||	is incomparable to	$x \parallel y$ means x is incomparable to y.
|...|	norm of; length of	$\| x \|$ means the norm of the element x of a normed vector space.
K	K	$\mathbf{K}$ means both $\mathbf{R}$ and $\mathbf{C}$: a statement containing $\mathbf{K}$ is true if either $\mathbf{R}$ or $\mathbf{C}$ is substituted for the $\mathbf{K}$.
$\sim$	is row equivalent	$A \sim B$ means that B can be generated by using a series of elementary row operations on A

About the Authors

Dariush Alimohammadi obtained a BA degree in Library and Information Sciences from Allame Tabataba'ee University, Tehran, in 1999, and an MA in Library and Information Sciences from Tehran University in 2003. Teaching in workshops, referring in conferences, collaborating on research projects, and seminar presentations are among his experiences. He is a member of the Editorial Board of *Library Philosophy and Practice*; the Editorial Review Board of *Informing Science: The International Journal of an Emerging Transdiscipline* and *Interdisciplinary Journal of Information, Knowledge, and Management*; and the International Editorial Advisory Board of *Indian Journal of Library and Information Science*. He has also reviewed papers for *International Journal of Information Science and Management*, *The Electronic Library*, and *Webology*, as well as the Persian LIS journals *Ettelashenasi = Informology, Faslname-ye Ketab = Book Quarterly* (The Quarterly Journal of the National Library and Archives of the Islamic Republic of Iran), and *Faslname-ye Olumo Fanavariye Ettelaat = Information Science and Technology Journal* (The Quarterly Journal of the Iranian Research Institute for Scientific Information and Documentation). Before joining to the Department of Library and Information Studies at the Tarbiat Moallem University as a Lecturer, he was involved in information services and systems design for about ten years. He has published more than forty English and Persian papers in national and international journals, two chapters, and one book in India. He is currently a member of directorate of the *Iranian Library and Information Science Association (ILISA)*.

Mary K. Bolin is Professor and Chair of Technical Services in the University Libraries at the University of Nebraska–Lincoln. She received a BA in Linguistics from the University of Nebraska–Lincoln in 1976, and an MSLS from the University of Kentucky in 1981. She earned a second Masters in English (linguistics) from the University of Idaho in 1999, and a PhD in Education from the University of Nebraska in 2007. Dr. Bolin is one of the founding editors of the peer-reviewed electronic journal *Library Philosophy and Practice*. She is also a Lecturer in the School of Library and Information Science at San José State University, where she has taught courses in information retrieval and metadata.

Preface

What is this book about?

This book is about Information Retrieval (IR), particularly Classical Information Retrieval (CIR). It looks at these topics through their mathematical roots. The mathematical bases of CIR are briefly reviewed, followed by the most important and interesting models of CIR, including Boolean, Vector Space, and Probabilistic.

Why this book?

Mathematics is a foundation and building block of all areas of knowledge. It particularly affects disciplines concerned with information organization, storage, retrieval, and exchange. Information is manipulated using computers, and computers have a mathematical basis. The word "computer" reveals this relationship. Students and practitioners of computer science, library and information science (LIS), and communications need a foundation in mathematics. IR, a subfield in all these disciplines, also needs mathematics as a common and formal language. Understanding CIR is not possible without basic mathematical knowledge.

Who is the audience for this book?

The primary goal of book is to create a context for understanding the principles of CIR by discussing its mathematical bases. This book can be helpful for LIS students who are studying IR but have no knowledge of mathematics. Weakness in math impairs the ability to understand current issues in IR. While LIS students are the main target of this book, it may be of interest to computer science and communications students as well.

How to read this book?

There are three ways of reading this book including:

1.   The obvious way is to read it from beginning to end, and in fact it has been designed for that. This approach satisfies the supposed LIS student;

2.   Another way is to read only the second part. Anybody with a good record in mathematics would be able to follow the second approach;

3.   The third way is to refer to the Appendix. It would be of benefit to those who are seeking admission to Computer and Information Science Schools with a mathematical approach. These schools mainly aim to execute programs and do research projects in the field of IR from the perspective of mathematics.

Acknowledgments

There are many people who have made the writing of this book possible:

First, I must express appreciation for my dear wife, Mahshid Sajjadi, for her infinite help and patience; then I would like to thank Mr. Iraj Saba and Dr. Abbas Horri, emeritus professors of LIS Departments at the Al-lame Tabataba'ee and Tehran Universities respectively, who always encouraged me to think, talk, and write logically;

I must also acknowledge Alistair Moffat, Amy N. Langville, Bob Plemmons, Christos Dimitrakakis, Claudio Gennaro, Danilo Montesi, Deborah MacPherson, Glynn Harmon, Ingemar Cox, J. Scott Olsson, Javed Aslam, Jesper Wiborg Schneider, Jimmy Lin, Jin Zhang, Juan C. Valle Lisboa, Kris Ven, Laurence A. F. Park, Ling-Hwei Chen, Marco Gori, Marc Sevaux, Michael W. Berry, Ming-Yen Lin, Mohammad Ahmadi, Peretz Shoval, Pertti Vakkari, Ralitsa Angelova, Shenghuo Zhu, Suh-Yin Lee, Tim Menzies, Tomas Subrt, Víctor Herrero-Solana, Vijay Raghavan, Vikas Sindhwani, Vladimir Pestov, and Vo Ngoc ANH who provided me with copies of their outstanding works;

Expert colleagues and the academic sphere they have created in the LIS Department of the Tarbiat Moallem University have been important in this achievement;

And finally, Mary K. Bolin, who introduced me to Paul Royster, director of the University of Nebraska–Lincoln Digital Commons, which is the publisher of this work.

Without these people, I would never have been able to realize such a big dream.

Part One

Roots

Chapter One

Set Theory

Introduction

Set is one of the basic concepts in mathematics. It is a group of objects, identified in a way that any object can be defined as part of the group or not. The following examples can be use to clarify:
- A set of one-digit numbers like 0, 1, 2, 3, 4, 5, 6, 7, 8, 9;
- The set of days in a week including Saturday, Sunday, Monday, Tuesday, Wednesday, Thursday, Friday;
- A set of countries started with C, such as Cambodia, Canada, Chile, China.

Each of the objects in a set is called an *element*. Every set is illustrated with a capital letter: A, B, ..., Z. In contrast, elements of a set are illustrated with lower case letters: a, b, ..., z. Sets can be categorized into two classes. The first class includes sets with a finite number of elements, and the second sets with an infinite number of elements. For example, the set of students in a school is a finite set, but the set of natural numbers is an infinite one. Visualizing sets is possible in the following ways:

1. *Naming elements.* There are three options:
   a. If we have a finite set, we enter elements in {} and place commas between them;

$$A = \{2, 4, 6, 8\}$$

   b. If there are too many elements to list, we list the first elements, followed by ellipses, and then the final element;

$$A = \{1, 2, 3, ..., 999\}$$

   c. If the set is infinite, we list the first elements, followed by ellipses;

$$A = \{1, 2, 3, ...\}$$

2.   *Using a Venn diagram* (Figure 1). A Venn diagram is a method for displaying relationships between subsets of a universal set. When elements of a set are illustrated in a circle, the Venn diagram is used as a geometrical interpretation. If A is a set of even numbers smaller than 10, then we can illustrate the geometrical diagram of even numbers.

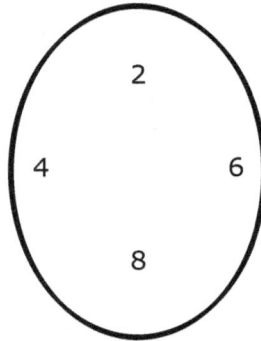

Figure 1 – Venn diagram of even numbers smaller than 10

3.   *Using mathematical symbols*. In this case, we can list elements of a set with a given property. For example, in $A = \{x : 1 < x < 9\}$, x is a variable. A variable, which can be represented by a letter or a sign, substitutes for an element of a set. In a more developed form, $p(x)$ expresses a given property of an element. In other words, elements of A have a common property in $p(x)$;

$$A = \{x : p(x)\}$$

Membership or non-membership in a set is illustrated with $\in$ and $\notin$ respectively.

Definition 1: A set with no elements is called empty or null set. The symbol of an empty set is $\emptyset$ or { }.

Definition 2: Every set may have several sub-sets (Figure 2).

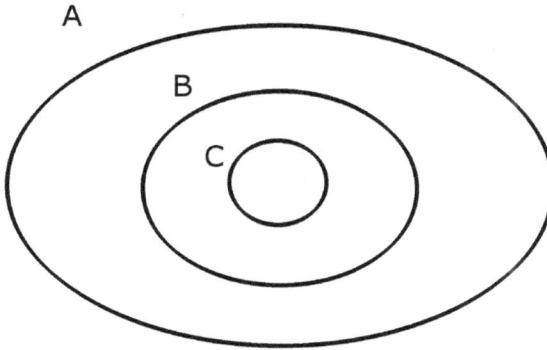

Figure 2 – A set with two sub-sets

In this figure, *A* as a set includes *B* and *C*. In the next level, *B* includes *C*. Inversely, *C* is the subset of *B*; and *B* is the subset of *A*. The essential condition for being a comprehensive set is that all elements of a subset exist in the set. Without this condition, the smaller set cannot be a subset of the larger one. As general rules, every set is the subset of itself, and the empty set is the subset of any set.

Definition 3: *A* and *B* are equal, if and only if, elements of *A* exist in *B* and elements of *B* exist in *A*;

$$A = B \text{ if and only if } A \in B \text{ and } B \in A$$

Sets may also overlap (Figure 3).

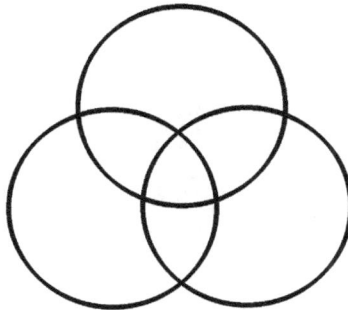

Figure 3 – Overlap among sets

Algebra of Sets

Algebra is used to describe the relationships among real numbers. The algebra of sets is used to describe the relationships among sets. Figure 3 illustrates overlap among sets. This concept brings us to the algebra of sets, which primarily follows the rules of Boolean algebra.

Definition 4: The union of A and B is a set that includes all the elements of A and B (Figure 4). The union is represented by the letter U;

$$A \cup B = \{x : x \in A \text{ or } x \in B\}$$

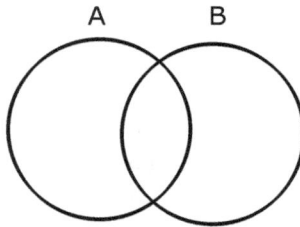

Figure 4 – Both sections are the union of A and B

Definition 5: The intersection of A and B is a set that includes only the common elements of both sets (Figure 5). The intersection is represented by ∩;

$$A \cap B = \{x : x \in A \text{ and } x \in B\}$$

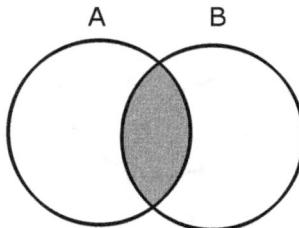

Figure 5 – The place where the circles overlap in the intersection of A and B

Definition 6: The complement of A and B is the elements of A that do not exist in B (Figure 6). The complement is represented by a minus sign;

$$A - B = \{x : x \in A \text{ and } x \notin B\}$$

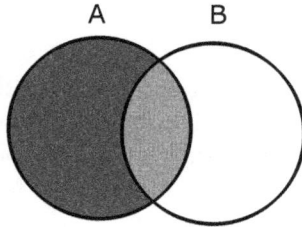

Figure 6 – The darker section (left) is the complement of A and B

Cartesian product

One of the methods for making new sets is the Cartesian product. Any two objects are a pair. We may have sorted and unsorted pairs. The sorted pair is showed with (a, b).

Definition 7: The Cartesian product of sets A and B is those sorted pairs (a, b) in which a is an element of A and b is an element of B. This is represented by $A \times B$;

$$A \times B = \{(a, b) : a \in A, b \in B\}$$

Chapter Two

Mathematical Logic

Introduction

The basis of mathematical logic is the proposition. A proposition is a statement that could be true or false. Its truth has not been proved. *If 4 is an even number, then 24 will be an even number too*, is an example of a proposition. A statement will be a proposition if the following conditions are met:

1. The proposition should be a sentence;
2. The proposition should be an informative sentence;
3. The proposition may be true or false;
4. The proposition has just one value.

The truth or untruth of a proposition is called its value and illustrated with T or F. A proposition is represented by p, q, r. Any proposition has one of two values (Figure 7):

p
T
F

Figure 7 – Two values of a proposition

Two equal propositions may have four relations including (Figure 8):

p	q
T	T
T	F
F	T
F	F

Figure 8 – Four relations of two propositions

Quantifier propositions

Quantifiers are symbols that stabilize propositions. There are three forms of quantifiers:

- *Universal quantifier*: stabilizes a general proposition;
- *Existential quantifier*: stabilizes a partial proposition; and
- *Restricted quantifier*: stabilizes a general proposition, but an empty one.

Multi-quantifier proposition

A quantifier proposition may have one variable, which is called a one-quantifier proposition. A quantifier proposition may also have several variables, which is called a multi-quantifier proposition.

Compound proposition

A compound proposition is composed of several simple propositions based on propositional relations. Propositional relations include:

- *Union relation*: This relation is used to unify two propositions. Values of such a relation are illustrated in Figure 9.

p	q	$p \wedge q$
T	T	T
T	F	F
F	T	F
F	F	F

Figure 9 – Values of a union relation

- *Subtraction relation*: This relation is used to subtract propositions from each other. Values of such a relation are illustrated by Figure 10.

p	q	$p \lor q$
T	T	T
T	F	T
F	T	T
F	F	F

Figure 10 - Values of a subtraction relation

- *Conditional relation*: This relation is used to show a given conditional relation between two propositions. Values of such a relation can be illustrated in Figure 11.

p	q	$p \Rightarrow q$
T	T	T
T	F	F
F	T	T
F	F	T

Figure 11 - Values of a conditional relation

- *Bi-conditional relation*: This relation is used to show a bi-conditional relation between two propositions. Values of such a relation can be illustrated in Figure 12.

p	q	$p \Leftrightarrow q$
T	T	T
T	F	F
F	T	F
F	F	T

Figure 12 - Values of a bi-conditional relation

Generally, propositions can be equivalent; every proposition may have its own negation; and conditional propositions with a true hypothesis and true consequence are called conditional theorems.

Methods of reasoning

Apart from the intuitive reasoning, there are two major reasoning styles:

- *Deductive*: we explain a proposition and then generalize it to a vast number of elements.
- *Inductive*: we survey a vast number of elements and conclude from them a general proposition.

Chapter Three

Number Systems

Introduction

We learn to use a decimal number system and learn to write, read, and operate on numbers with facility. The decimal system is not the only available number system. Suppose that we have thirteen asterisks (*), displayed as follows:

$$* \ * \ * \ * \ * \ * \ *$$
$$* \ * \ * \ * \ * \ *$$

If we use the decimal system, we will have a set of ten elements and another set of three elements. The result is one of the oldest methods of categorizing elements. Ancient people categorized things in decimal sets so that every ten elements made one set. In our example, ten is basis for categorization. Categorizing elements based on another numbers is also possible; however, $n \geq 2$ is the basis of any number system. The following are some popular number systems:

- **Binary system**
  This system has two symbols: 0, 1. It categorizes into sets of two.

- **Octal system**
  This system has eight symbols: 0, 1, 2, 3, 4, 5, 6, 7. It categorizes into sets of eight.

- **Decimal system**
  This system has ten symbols: 0, 1, 2, 3, 4, 5, 6, 7, 8, 9. It categorizes into sets of 10. Decile and percentile are two common examples.

- **Hexadecimal system**
  This system has sixteen symbols: 0, 1, 2, 3, 4, 5, 6, 7, 8, 9. For the next six numbers, English letters from A to F are used. It categorizes into sets of 16.

Transformation among number systems

Transformation among various number systems is often needed to do calculations. We can transform a number from decimal system to a non-decimal system; from a non-decimal to a decimal system; and from a non-decimal system to another non-decimal system. Such a process is possible because of a common basis, i.e. $n \geq 2$. All number systems have a common basis in the number 2, an even number that can be changed simply to other even numbers. Furthermore, number systems can be added, subtracted, multiplied, and divided through algebraic operations.

Rounding-off error

Since the binary system just has two symbols (0, 1), it is an efficient system for computer applications. There are two main advantages for the binary system:

- Physical systems have mainly two conditions. For this reason, a binary approach works well for them. In an electric circuit, for example, presence and absence of voltage can be shown using 1 and 0.

- Since there are few numerals, few rules are needed for addition and multiplication.

There is one major disadvantage for binary number systems. One is forced to use multiple numerals, even for relatively small numbers.

- These characteristics of binary systems lead to rounding errors, because there is always a numeral smaller or larger than 5 at the end of decimal numbers.
  If the $(n + 1)$ numeral is smaller than 5, we cancel it. For example, rounding-off 23.341752 to two decimal numerals (places) results in 23.34. The 4 is not changed, because the next numeral is 1, which is smaller than 5, and rounding up occurs at 5.
  If the $(n + 1)$ numeral is larger than 5, we add one numeral to

the previous one. For example, rounding-off 23.341752 to three decimal numerals (places) results in 23.342. The "1" has 1 added to it, because the next numeral is 7, which is bigger than 5 and changes the previous numeral from 1 to 2.

Chapter Four

Combinatorics

Introduction

There are several principles for counting. Counting indicates how many possible arrangements are there for sorting objects or elements of a set. Counting methods are sometimes called "methods of combinatorics." We need a way to count and display all possible states of a set of elements. *Multiplication* and *addition* are important principles of counting. The *inclusion-and-exclusion* principle, *inversion* principle, and *Stirling numbers* are other counting principles.

Determining the number of possible states for an occurrence is one of the main challenges in some probabilistic problems. In most cases, there is no need to determine all of states, but only to determine the number of elements. Suppose that you go to have lunch in a restaurant. You take the menu and make a decision to order plain boiled rice, salad, and a beverage. There are two types of plain boiled rice, two types of salad, and two types of beverages. What are all the possible combinations that could be ordered? (Figure 13)

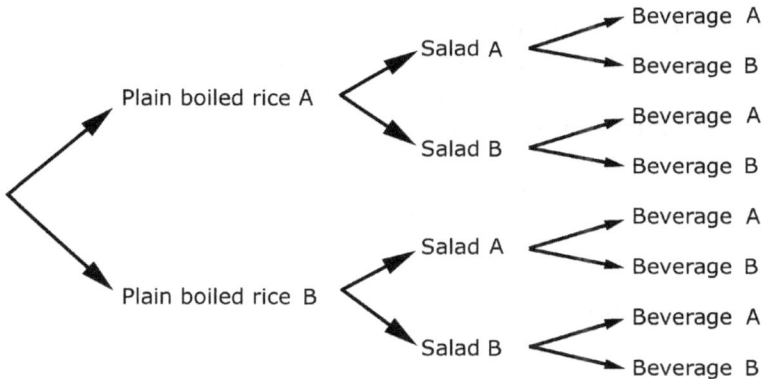

Figure 13 – Lunch combinations

You have two choices with any plain boiled rice; and two choices with any salad. Therefore, you will have a total of 8 choices;

$$2 \times 2 \times 2 = 8$$

As another example, suppose that you have two coats and four pairs of pants. In how many combination can you wear each coat and pair of pants? Obviously, there are two choices for the coat and four choices for the pants. Therefore, you will have a total of 8 choices;

$$2 \times 4 = 8$$

Multiplication Principle

If you are able to do one operation in m ways and another operation in n ways, then you would have $m \times n$ choices to do both operations simultaneously.

Addition principle

In case of:

- $E_1, E_2, ..., E_r$ events, with r indicating the number;
- E_i event could happen in n_i forms; and
- Neither the first event nor the second happened simultaneously;
- The total number of ways in which one of the events could happen equals:

$$n_1 + n_2 + ... + n_r$$

Transformation

If you have three books, how many ways can you shelve them? If we suppose three places for shelving, there are three possibilities for shelving the first book, two for the second book, and just one for the third book (Figure 14).

The first place	The second place	The third place
3	2	1

Figure 14 – The possibilities for shelving three books in three places

According to the multiplication principle, the number of possible states is equal to:

$$3 \times 2 \times 1 = 6$$

Arrangement

Suppose that you have n different objects and want to arrange r objects out of them beside each other ($r \leq n$). In how many ways can you do that? As mentioned, you set r places. The first place is filled out with n possibilities, the next place with $n - 1$ possibilities; and so on (Figure 15).

The first place	The second place	The r-th place
n	$n - 1$	$n - r + 1$

Figure 15 – The possibilities for arranging n objects in r places

Therefore, according to the multiplication principle the number of possible states is equal to:

$$n \times (n - 1) \times (n - 2) \times \ldots \times (n - r + 1)$$

Combination

If the arrangement is not important in selecting r objects out of n, it is called "combination."

Inclusion-and-exclusion principle

X is a finite set. The number of elements in this set is illustrated by $n(X)$. If A and B are finite sets and $A \cap B = \emptyset$, then $n(A \cup B) = n(A) + n\,B)$. If the intersection of A and B is not empty, use $n(A) + n(B)$ to calculate the number of elements of $A \cup B$. In this addition, common elements of A and B (their inclusion) have been calculated twice. The first time was when elements of A were enumerated, and the second time when elements of B were enumerated. For this reason, one of the two calculations should be omitted (Figure 16).

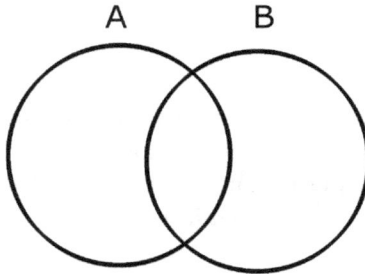

Figure 16 – The overlap between the two circles is the intersection of A and B

For two finite sets, therefore, the inclusion-and-exclusion principle is expressed as follows:

$$n(A \cup B) = n(A) + n(B) - n(A \cap B)$$

Inversion principle

Suppose that you intend to arrange numbers 1, 2, 3, …, 10 in a manner that every number stands in its own place. These arrangements are called inversions and are illustrated with d_n. d is a symbol for inversion and n is the number of elements of the set.

Stirling numbers

There are two kinds of Stirling numbers:

1.   The integers $s(n, k)$ generated by the recursive definition:

$$s(0, 0) = 1;$$
$$s(n, 0) = 0 \ (n > 0), \ and,$$

for $0 < k < n,$

$$s(n + 1, k) = s(n, k - 1) - ns(n, k);$$

and

2.   Numbers of the second kind, n. The natural numbers $S(n, k)$ generated by the recursive definition:

$$S(n, n) = 1 \ (n > 0),$$
$$S(n, 0) = 0 \ (n \geq 0), \ and,$$

for $0 < k < n,$

$$S(n + 1, k) = S(n, k - 1) + kS(n, k)$$

Chapter Five

Other Areas of Basic Mathematics

Generating Function

Suppose that you want to fill five boxes with three balls. How many ways are there for doing that? In our discussion of combinatorics, we learned that by using the multiplication principle, we can calculate an answer to this problem. Suppose that you are going to put balls in boxes separated from each other by vertical lines. The first and the last lines do not move. Between them are four moving lines and three boxes (Figure 17). If we show each place n with $n + 1$ vertical line, assuming the first and last lines constant, then filling out n places with r balls is equal to $n - 1$ vertical line.

*			* *	

Figure 17 – Filling boxes with three balls

Exponential function

Functions can be categorized in two general groups. The above-mentioned function is a general one. The general function does not need arrangement. In another type of function, the exponential, arrangement is necessary.

Recursive Relations

Follow the values sequence a_0, a_1, a_2, a_3, ... This simple sequence is called a *recursive relation* or *difference equation*. If $a_0 = 1$ and $a_1 = 2$, then $a_0 + a_1 = 3$ and $a_1 + a_2 = 5$. Recursive relations can be applied in counting and predicting possible conditions of a given set of elements. The most famous example of the application of the recursive relation is the *Towers of Hanoi* (Figure 18).

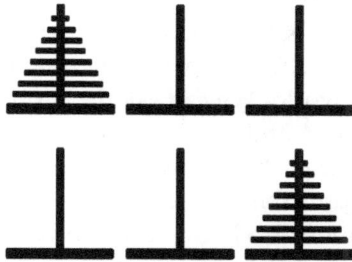

Figure 18 - The Towers of Hanoi

Suppose that you must transport disks around the left shaft to the right shaft. The transportation must be done without putting the bigger disk next the smaller disk. How many steps does it take? The answer is three steps. The best solution is to use the middle shaft. Transport all the disks except the biggest one to the middle shaft. In this phase, we will have 1 disk on the left shaft and $n - 1$ disks on the middle shaft. Then, transport the biggest disk to the right shaft. We still have 1 and $n - 1$. In the third phase, transport $n - 1$ disks to the right shaft just above the biggest disk.

Probability Theory

Probability theory is the mathematical theory of the notions of chance, randomness, and phenomena. The term probability addresses the possibility of occurrence of an unpredictable event. In descriptive statistics, we extract some parameters and characteristics of a given population. If we intend to generalize the parameters and characteristics to the whole population, we must use probability theory. In probability theory, several terms are very important, including *test, random test, sample space, event,* and *mathematical expectation.* In general, the probability of the occurrence of an unpredictable event may influence subsequent events. Tools like charts and diagrams can be used to show the probability of occurrence of events. *Tests* are the basis of any probabilistic calculation. The test may be *random* or not. However, as an *event,* it happens in a *sample space* and the *mathematical expectation* is the ideal or the optimum point of our probabilistic viewpoint (Manning, Raghavan, and Schutze 2008).

The variable A, for example, represents an event (a subset of the set of possible outcomes). Equivalently, we can represent the subset via a *random variable*, random which is a function from outcomes to real numbers; the sub-variable set is the domain over which the random variable A has a particular value. Often we will not know with certainty whether an event is true in the world. We can ask the probability of the event $0 \leq P(A) \leq 1$. For two events A and B, the joint event of both events occurring is described by the joint probability $P(A, B)$. The conditional probability $P(A \mid B)$ expresses the probability of event A given that event B occurred. The fundamental relationship between chain rule joint and conditional probabilities is given by the chain rule (Manning, Raghavan, and Schutze 2008):

$$P(A, B) = P(A \cap B) = P(A \mid B)\,P(B) = P(B \mid A)\,P(A)$$

Without making any assumptions, the probability of a joint event equals the probability of one of the events multiplied by the probability of the other event conditioned on knowing the first event happened.

Writing $P(\bar{A})$ for the probability of the complement of an event A, we similarly have:

$$P(\bar{A}, B) = P(B \mid \bar{A})\,P(\bar{A})$$

Probability theory also has a *partition rule*, which says that if an event B can be divided into an exhaustive set of disjoint sub-cases, then the probability of B is the sum of the probabilities of the sub-cases. A special case of this rule gives that:

$$P(B) = P(A, B) + P(\bar{A}, B)$$

From these we can derive *Bayes' rule* for inverting conditional probabilities:

$$P(A \mid B) = \frac{P(B \mid A)\,P(A)}{P(B)} = \left[\frac{P(B \mid A)}{\sum_{X \in \{A, \bar{A}\}} P(B \mid X)\,P(X)} \right] P(A)$$

This equation can also be thought of as a way of updating probabilities. We start off with an initial estimate of how likely the event A is when we do not have any other information; this is the *prior probability P(A)*. Bayes' rule of probability lets us derive a *posterior probability P (A | B)* after having seen the evidence *B*, based on the *likelihood* of *B* occurring in the two cases that *A* does or does not hold. Finally, it is often useful to talk about the *odds* of an event, which provide a kind of multiplier for how probabilities change (Manning, Raghavan, and Schutze 2008):

$$Odds: \quad O(A) = \frac{P(A)}{P(\bar{A})} = \frac{P(A)}{1 - P(A)}$$

Graph

A graph consists of a number of vertices (or points or nodes), some of which are joined by edges. The edge joining the vertex *u* and the vertex *v* may be denoted by (u, v) or (v, u). The vertex-set that is the set of vertices of a graph *G* may be denoted by *V (G)* and the edge-set by *E (G)*. In general, a graph may have more than one edge joining a pair of vertices. When this occurs, these edges are called multiple edges. Also, a graph may have loops – a loop is an edge that joins a vertex to itself.

In another words, a **graph** is a geometric representation of a relationship between numbers, usually in a rectangular coordinate system. The construction of a visual representation of a graph is called **graphing**. **Composition** is a method of graphing which consists of writing the given function as the sum of several functions whose graphs are easier to draw, plotting each of these functions, then adding the corresponding ordinates. Graphs are used primarily to illustrate a relationship among elements in a multi-dimensional space.

Algorithms

Algorithms are instructions for carrying out a series of logical procedural steps in a specified order. The term is now used especially in computing and related disciplines. In the *Towers of Hanoi* example, an algorithm was used to transport disks. Another example of the application of an algorithm in daily life is the use of an ATM to do banking.

The following commands constitute an algorithm for using an ATM:
- Insert your card;

- Enter your code;
- Select a payment option;
- Get your card;
- Get your payment;
- Get the receipt.

An algorithm is a finite set of commands, which must be unambiguous. Several algorithms might be used to solve a given problem. Algorithms may be combined to create one sophisticated algorithm to solve large or complex problems. Algorithms can be classified according to their degree of sophistication. Algorithms should be written in a way takes the shortest route to the result.

Networks

A network is a digraph in which every arc is assigned a weight (some non-negative number). In some applications, something may be thought of as flowing or being transported between the vertices of a network, with the weight of each arc giving its capacity. In other cases, the vertices of a network may represent steps in a process and the weight of the arc joining u and v may be the time that must elapse between step u and step v.

Part 2

Applications

Chapter Six

Basics of Classical Information Retrieval

Background

van Rijsbergen (1979) states that,

> Since the 1940s, the problem of information storage and retrieval has attracted increasing attention. Vast amounts of information need accurate and speedy access. One effect of this is that relevant information gets ignored since it is never uncovered, which in turn leads to much duplication of work and effort. With the advent of computers, a great deal of thought has been given to using them to provide rapid and intelligent retrieval systems. In libraries, many of which certainly have an information storage and retrieval problem, some of the more mundane tasks, such as cataloguing and general administration, have successfully been taken over by computers. However, the problem of effective retrieval remains largely unsolved. (p. 3)

Kowalski (1997) observes that,

> The first Information Retrieval Systems originated with the need to organize information in central repositories (e.g., libraries). Catalogues were created to facilitate the identification and retrieval of items. As computers became commercially available, they were obvious candidates for the storage and retrieval of text. Early introduction of Database Management Systems (DBMSs) provided an ideal platform for electronic manipulation of the indexes to information. Libraries followed the paradigm of their catalogs and references by migrating to the format and organization of their hardcopy information references into structured databases. These remain as a primary mechanism for researching sources of needed information and play a major role in available Information Retrieval Systems.

van Risjbergen (1979) continues,

> When high speed computers became available for non-numerical work, many thought that a computer would be able to "read"

an entire document collection to extract the relevant documents. It soon became apparent that using the natural language text of a document not only caused input and storage problems (it still does) but also left unsolved the intellectual problem of characterizing the document content. ... [A]utomatic characterization in which the software attempts to duplicate the human process of "reading" is a very sticky problem indeed. More specifically, "reading" involves attempting to extract information, both syntactic and semantic, from the text and using it to decide whether each document is relevant or not to a particular request. The difficulty is not only knowing how to extract the information but also how to use it to decide relevance. (pp. 3–4)

What is and what does an Information Retrieval System?

Manning, Raghavan, and Schütze (2008) observe that,

Information retrieval (IR) is finding material (usually documents) of an unstructured nature (usually text) that satisfies an information need from within large collections (usually stored on computers). As defined in this way, information retrieval used to be an activity that only a few people engaged in: reference librarians, paralegals, and similar professional searchers. Now the world has changed, and hundreds of millions of people engage in information retrieval every day when they use a web search engine.

Kowalski (1997) continues,

An Information Retrieval System is a system that is capable of storage, retrieval, and maintenance of information. Information in this context can be composed of text (including numeric and date data), images, audio, video and other multi-media objects. Although the form of an object in an Information Retrieval System is diverse, the text aspect has been the only data type that lent itself to full functional processing. The other data types have been treated as highly informative sources, but are primarily linked for retrieval based upon search of the text.

A major part of an Information Retrieval System is a software program that facilitates finding the information the user needs. The system may use standard computer hardware or specialized hardware to support the search sub-function and to convert non-textual sources to a searchable media (e.g., transcription of audio to text). The gauge of success of an information system is how well it can minimize the overhead for a user to find the needed information. Overhead from a user's perspective is the time required to find the information needed, excluding the time for actually reading the relevant data. Thus search composition, search execution, and reading non-relevant items are all aspects of information retrieval overhead.

Van Rijsbergen (1979) explains,

> In an overview, an information retrieval system has three components: input, processor and output. The main problem here [with the input] is to obtain a representation of each document and query suitable for a computer to use. ... A *document representative* could, for example, be a list of extracted words considered to be significant. Rather than have the computer process the natural language, an alternative approach is to have an artificial language within which all queries and documents can be formulated. ... Secondly, the processor, that part of the retrieval system concerned with the retrieval process. The process may involve structuring the information in some appropriate way, such as classifying it. It will also involve performing the actual retrieval function that is, executing the search strategy in response to a query. ... Finally, we come to the output, which is usually a set of citations or document numbers. In an operational system the story ends here. However, in an experimental system it leaves the evaluation to be done. (pp. 4-5)

> In principle, information storage and retrieval is simple. Suppose there is a store of documents and a person (user of the store) formulates a question (request or query) to which the answer is a set of documents satisfying the information need expressed by his question. He can obtain the set by reading all the documents in the store, retaining the relevant documents and discarding all the others. In a sense, this constitutes "perfect" retrieval. This solution is obviously impracticable. A user either does not have the time or does not wish to spend the time reading the entire document collection, apart from the fact that it may be physically impossible for him to do so. (p. 3)

Logical models of Information Retrieval represent documents and queries as logical formulas, and apply some form of inference provided by the logic to decide relevance. The simplest approach is to model the relevance test using the logical entailment $d \to q$, where d and q are logical representations of a document and a query, respectively.

The Concept of Similarity

Park, Ramamohanaro, and Palanaswami (2005) observe that,

> Many current information retrieval systems are built around a similarity function. This function takes a query and a document as its arguments and generates a single score which represents the relevance of the query to the document. Current popular retrieval models, namely, Vector Space and Probabilistic models, base their similarity function on the hypothesis that a document is more likely to be relevant to a query if it contains more occurrences of the query terms.

This implies that the similarity functions need the count of occurrences of each of the terms, and ignore any other information in the document. Such a vision sets forth the similarity information retrieval. Similarity Information Retrieval (SIR) is now defined as a special case of Classical Information Retrieval (CIR). SIR is equivalent to the classical vector IR; hence the latter is a special case of CIR. The vector IR has two (traditional) particular cases: binary and non-binary. These can easily be defined as being two special cases of SIR as follows:

- Binary Similarity Information Retrieval (BSIR) is a SIR $< D, R >$ with $S = \{0, 1\}$;
- Non-Binary Similarity Information Retrieval (NBSIR) is a SIR $< D, R >$ with $S = [0, 1]$.

Information Retrieval Models

Dominich (2001) says that, "the basic theoretical model types of IR reflect, on the one hand, the complexity and interdisciplinarity of IR in general and of IR modeling in particular and, on the other hand, the modeling difficulty of the most critical parameter that is relevance."

The basic theoretical model types of IR differ from each other in the way objects (e.g., documents, images) are represented and retrieval is defined. Mathematical IR represents objects as strings of numbers (traditionally called vectors), and defines retrieval using numeric relationships (e.g., a distance meant to express a likeness) between vectors. Logical IR assumes objects as representations (e.g., collections of sentences, situations) and retrieval as an uncertain logical inference. Interaction IR views objects as flexibly interconnected active elements, and retrieval as memories of elements recalled by a query. Artificial intelligence (AI) IR conceives objects as a Knowledge Base (KB), and retrieval as some reasoning or as neurons and retrieval as a spreading of activation.

Different generalizations of the basic mathematical model types yielded the idea of creating a unified mathematical foundation for them. Because these generalizations shifted the interpretation of conditional probabilities toward logic, Logical IR models have been elaborated with the aim to contain the basic mathematical models, too, as special cases.

However, in CIR, the content analysis of the documents is used to decide whether a document is relevant to a particular query. A general concept of CIR is defined first from which the vector and probabilistic IR models can be formally obtained as special cases. CIR might be seen as a superstructure with a parameter, from which the vector model is obtained taking a particular value for the parameter, whereas taking another value for it, the probabilistic model is obtained.

Chapter Seven:

Boolean Model of Information Retrieval

Introduction

Boolean retrieval poses queries in Boolean expressions of terms. Terms are combined with the operators AND, OR, and NOT. The model views each document as just a set of words (Manning, Raghavan, and Schütze 2008).

Boolean Logic as the Base of Boolean IR

Boolean logic allows a user to relate multiple concepts to describe needed information. Boolean functions apply to text found anywhere within a document. The Boolean operators are AND, OR, and NOT. These operations are use set intersection, union, and complement. A few systems use the concept "exclusive or" that may also be called exclusive disjunction (symbolized by XOR or EOR). Putting portions of the search statement in parentheses specifies the order of operations. Without parentheses, the default precedence is NOT, AND, OR (Manning, Raghavan, and Schütze 2008).

One kind of of Boolean search is "M of N" logic. The user lists search terms and searches for any item that contains a subset of the terms; for example, any item containing two of the following terms: "AA", "BB", and "CC". This can be made into a Boolean search that does an AND search between all combinations of two terms and searches the results using OR: (Manning, Raghavan, and Schütze 2008)

(AA AND BB) or (AA AND CC) or (BB AND CC)

Most systems allow Boolean searching. Little attention has been given to putting Boolean search functions and weighted retrieval techniques into one search result.

Wavelet Model

A document may be more relevant if query terms appear in a pattern throughout the text. This could be examined using proximity or by mapping term positions to another domain. "Wavelet transforms" (Park, Ramamohanarao, and Palaniswami 2005) allow the discovery of patterns. The wavelet transform breaks a signal into wavelets of different scale and position. This allows the analysis of the signal at different frequencies, to identify spikes in the signal. A term that is frequent but scattered can be represented by a low-frequency wavelet, while a term that appears once is a high-frequency positional wavelet (Park, Ramamohanarao, and Palaniswami 2005).

A wavelet is described by a function $\psi \in L^2(\mathbb{R})$ (where $L^2(\mathbb{R})$ is the set of functions $f(t)$ which satisfy $\int |f(t)|^2 \, dt < \infty$) with a zero average and norm of 1. A wavelet can be scaled and translated by adjusting the parameters s and u, respectively.

$$\psi_{u,s}(t) = \frac{1}{\sqrt{s}} \, \psi \left(\frac{t - u}{s} \right)$$

The scaling factor keeps the norm equal to one for all s and u. The wavelet transform of $f \in L^2(\mathbb{R})$ at time u and scale s is

$$W(u, s) = (f, \psi_{u,s}) = \int_{-\infty}^{+\infty} f(t) \frac{1}{\sqrt{s}} \psi* \left(\frac{t-u}{s} \right) dt,$$

where $\psi*$ is the complex conjugate of ψ.

A wavelet function, $\psi u,s(t)$, requires a scaling function, $\varphi u,s(t) \in Vn$. The scaling function must satisfy the property

$$\ldots \subset V_{n+1} \subset V_n \subset V_{n-1} \ldots,$$

where the set of $\varphi_{u,s}(t)$ for all u is a basis of V_n ($s = 2^n$ for dyadic scaling), and $\bigcup_{n \in \mathbb{Z}} V_n = L^2(\mathbb{R})$. Each set of scaling functions can be shown in terms of its subset of scaled scaling functions:

$$V_{n-1} = V_n \cup W_{n-1}, \quad V_n \perp W_{n-1},$$

where $\perp$ implies orthogonality. The set of functions W_n, satisfies the following properties:

$$\bigcup_{n \in \mathbb{Z}} V_n = \mathbf{L}^2(\mathbb{R}), \quad \bigcap_{n \in \mathbb{Z}} W_n = \emptyset$$

Therefore the set of functions Wn for all n is a basis for $\mathbf{L}^2(\mathbb{R})$. This set W_n is the set of shifted wavelet functions at resolution n (Park, Ramamohanarao, and Palaniswami 2005).

Spectral-Based Document Retrieval

Spectral-based retrieval considers occurrence patterns of query terms. Query terms with similar positional patterns make their documents more relevant than those without such patterns (Park, Ramamohanarao, and Palaniswami 2005).

While vector space retrieval calculates relevance based on the occurrence of query terms, and proximity models calculate it based on the proximity of query terms to each other, proximity searching uses more information from the document to calculate the score, but takes more time. Spectral-based retrieval overcomes this problem by comparing terms in their spectral, rather than their spatial, domain. A term signal for each query term is created, and the term signals are converted into term spectra using a spectral transform (Park, Ramamohanarao, and Palaniswami 2005).

1.   Term Signals

A term signal shows the occurrence of a particular term in a particular section of a document. The term signal for term t in document d is represented by

$$\tilde{f}_{d,t} = [f_{d,t,0} \ f_{d,t,1} \ \cdots \ f_{d,t,B-1}]$$

where $f_{d,t,b}$ is the value of the signal component. If there are B signal components and D terms in the document, the value of the b-th component is calculated by counting occurrences of term t between the bD/B-th word in the document and the $\{(b + 1)D/B - 1\}$-th word in the document. Therefore, if $B = 8$, $f_{d,t,0}$ would contain the number of times term t occurred in the first eighth of document d. If $B = 1$, $f_{d,t,0}$ would contain the count of term t throughout the whole document (Park, Ramamohanarao, and Palaniswami 2005).

2.   Term Signal Weights

Weighting schemes in vector space and probabilistic methods are used to reduce the impact of certain document and term properties on the document score. These exist in term signals as well, and weighting is used to reduced their impact.

Each component of a term signal represents a portion of the document; existing document weighting schemes can be used to weight each of the term signal components.

The document weights used were

$$w_{d,t,b} = \frac{1 + log(f_{d,t,b})}{(1-s) + sW_d/W_d}$$

$$w_{d,t,b} = \frac{f_{d,t,b}}{f_{d,b,t} + \tau_d/\overline{\tau}_d}$$

$$w_{d,t,b} = \frac{1 + log(f_{d,t,b})}{(1-s) + sW_d/W_d}$$

$$w_{d,t,b} = \frac{(1 + log(f_{d,t,b}))/(1 + log(\tilde{f}_{d,t}))}{((1-s) + s\tau_d/\overline{\tau}_d)}$$

where $f_{d,t,b}$ is the b-th component of the t-th term in the d-th document, $\tilde{f}_{d,t}$ is the average term count for document d, W_d is the document vector l_2 norm, τ_d and $\overline{\tau}_d$ the number of unique terms in document d and the average unique terms, respectively, and s is a slope parameter (Park, Ramamohanarao, and Palaniswami 2005).

3.   Term Spectra

To compare query term signals to obtain a document score, component b of each term or different components in different terms could be compared. The former would reduce to passage retrieval, while the latter is a form of proximity measure. Term signal positions are not compared, but their patterns are. This can be done by examining their wavelet spectrum

$$\tilde{\zeta}_{d,t} = [\zeta_{d,t,0} \ \zeta_{d,t,1} \ \cdots \ \zeta_{d,t,B-1}]$$

where $\zeta_{d,t,b} = H_{d,t,b} \exp(i\theta_{d,t,b})$ is the b-th spectral component with magnitude $H_{d,t,b}$ and phase $\theta_{d,t,b}$. These transforms extract frequency information, but they focus on the signal as a whole. The wavelet transform focuses on parts of the signal at different resolutions. Frequency information is extracted from parts of the document, which results in frequency and position information. The resulting term spectrum contains orthogonal components, meaning that there is no need to compare spectral components (Park, Ramamohanarao, and Palaniswami 2005).

Chapter Eight

Vector Space Model of Information Retrieval

Introduction

The Vector Space Model (VSM) is a classical model that has been used to process texts for about forty years. In this model, documents and queries are each mapped to a point based on frequency, using Euclidean geometry (Góth and Skrop 2005). The term "vector IR" can be misleading. In vector IR, objects are represented as strings of real numbers, as if they were mathematical vectors, although they not necessarily form a vector space (Dominich 2001). Vector space represents documents and queries, with one dimension for each term. The number of appearances of each term in the document forms a dimension of the document vector. The "similarity function" applies weights to vectors and assigns a relevance score (Park, Ramamohanarao, and Palaniswami 2005).

Queries and documents are represented as strings of numbers: vectors. Every number represents the degree to which a term characterizes a document, based on frequency. Thus, a similarity measure expresses the likeness between queries and documents. The similarity measure is generally normalized (has a value between 0 and 1). This property is called *normalization*. The value of the similarity measure does not depend on the order in which queries and documents are compared (*symmetry* or *commutativity*). Finally, the similarity measure is maximal (equal to 1) when the vectors of queries and documents are identical. This is the property of *reflexivity* (Dominich 2000).

Vector Space Model

A formal definition of vector IR is needed:

Definition of Vector IR (VIR). *Let D be a set of objects (documents). A function o: D × D → [0, 1] is called a similarity if the following three properties (a) through (c) hold:*

(*a*) $0 \leq (a, b) \leq 1. \quad \forall \, a, b \in D$,   *normalization*

(*b*) $\sigma(a, b) = \sigma(b, a). \quad \forall \, a, b \in D$,   *symmetry or commutativity*

(*c*) $a = b \Rightarrow \sigma(a, b) = 1. \quad a, b \in D$,   *reflexivity*

Let $q \in D$ be a query, and $t \in \mathbb{R}$ be a real threshold value. The set $\mathbf{R}(q)$ of retrieved documents in response to query q is defined as follows:

$$\mathbf{R}(q) = \{d \in D \mid \sigma(d, q) > T, \, T \in \mathbb{R}\}$$

As a broader description, VSM can be introduced as (Góth and Skrop 2005):

Given documents D_j, $j = 1, \ldots, m \in \mathbb{N}$ ($\mathbb{N}$ denotes the set of natural numbers), and terms t_i, $i = 1, \ldots, n \in \mathbb{N}$. Using the VSM, every document D_j is assigned a vector $\mathbf{w}_j = (w_{ij})_{i\,=1, \ldots,\, n}$ of weights, where $w_{ij} \in \mathbb{R}$ ($\mathbb{R}$ denotes the set of real numbers) denotes the weight of term t_i for document D_j. The matrix $W = (w_{ij})_{n \times m}$ is called the term-by-document matrix. The general form of a weighting scheme is as follows:

$$w_{ij} = local\_weight_{ij} \times global\_weight_i \times normalization_j = l_{ij} \times g_i \times n_j$$

Let Q denote a user's *query*, and $\mathbf{q} = (qi)\, i = 1,\ldots, n$ the corresponding query weight vector. The vectors $\mathbf{w}_j$ and $\mathbf{q}$ belong to the E_n Euclidean orthonormal space, in which the weights $\mathbf{w}_j$ and $\mathbf{q}$ are Cartesian coordinates (of points corresponding to document D_j and query Q). Each term t_i is assigned to an axis x_i. All the axes intersect at one common point O (the origin). They are pair-wise perpendicular to each other in the origin, and the weight w_{ij} corresponds to a point on the axis x_i (with one point for each document D_j). Thus, every document D_j is represented by a vector w_j, which defines a point in the space E_n. The *relevance* of document D_j relative to query Q is given by the value of a *similarity measure* σ $(\mathbf{w}_j, \mathbf{q})$, whose general form is:

$$\sigma\,(\mathbf{w}_j, \mathbf{q}) = \mathbf{w}_j\,\mathbf{q}\,/\,\Delta$$

where $\mathbf{w}_j\,\mathbf{q}$ denotes the inner $-$ or dot $-$ product of the vectors $\mathbf{w}_j$ and $\mathbf{q}$. Several similarity measures have been proposed, such as the Dot product, Cosine, and the coefficients (overlap, Dice's, Jaccard's). The following uses the Cosine measure. Its explicit formula is obtained for

$\Delta = \|\mathbf{w}_j\| \cdot \|\mathbf{q}\|$ ($\| \cdot \|$ denotes the Euclidean norm of a vector). The concept of a *σ-space* is a formal generalization of the VSM to emphasize similarity measures. A set D of objects with a symmetric and reflexive similarity measure, i.e.,

$$\sigma: D \times D \rightarrow R$$

Symmetry: $\sigma(a, b) = \sigma(b, a)$, $\forall a, b \in D$; i.e., the order in which the query and the document are considered is irrelevant;

Reflexivity: $a = b \Rightarrow \sigma(a, b) = \kappa$; i.e., the similarity measure is equal to a maximal value κ if the query and the document are exactly the same; but the reverse is not necessarily true. For example, if σ is normalized, κ may be taken as being equal to 1. It is referred to as a *σ-space*.

A document is defined by n independent features or attributes, used to describe the subject(s) of the document. In most cases, these are keywords from the title, abstract, or full-text from the document.

$$d_i = (a_{i1}, a_{i2}, ..., a_{ij}, ..., a_{in})$$

d_i is a document, a_{ij} is a feature describing the document. Its value or weight reflects the importance of this feature a_{ij} to document d_i, valid value of a_{ij} ranges from 0 to infinity, and n is the number of features or the dimensionality of the vector space. R^n denotes a vector with n dimensionality. A vector corresponds to a visible point in a low dimensional space (for instance, a two or three dimensional space), or an invisible point in a higher dimensional space. For a linear vector space, if d_1, d_2, and $d_3 \in R^n$, c is a constant, the following equations always hold true.

$$(d_1 + d_2) \times c = d_1 \times c + d_2 \times c$$
$$d_1 + d_2 = d_2 + d_1$$
$$(d_1 + d_2) + d_3 = d_1 + (d_2 + d_3)$$

A matrix is a table or rectangular array of elements arranged in rows and columns. A document-term matrix is a group of document vectors. The rows and columns are documents and features respectively.

Where a_{ij} is the weight of document d_i for feature j, m is the number of the documents in a document collection. A query can also be defined as a vector. Where q_j is the weight of feature j and its value is dependent upon a user's information need, n is the number of unique features which should be equal to n in the equation for d_i above. Query representation structure is the same as a document representation structure, which makes various calculations between a document and a query possible.

$$q = (q_1, q_2, ..., q_j, ..., q_n)$$

The number of unique features (n) in a document-term matrix can be large, because the features are unique indexing terms used in a document collection. As the number of documents indexed in a collection increases, the number of the features (n) also increases. The relationship between the number of documents and the number of features (n) is not simply linear, however. When the number of documents indexed in a collection reaches to a certain level, the number of features (n) stays stable.

Looking at each of the documents in the matrix, we find that the number of non-zero features relatively small compared to the number of features (n). The number of non-zero features is affected by indexing practices and length of the document. As a result, the document-term matrix is a sparse matrix where most of its elements are 0. (Zhang 2008)

The strengths of the vector space model are summarized as follows:

- The vector-based structure can represent an object with multiple attributes;
- Weights can be assigned to indexing terms to distinguish their significance;
- Similarly, weights can also be assigned to query terms, expressing users' needs in a more accurate and flexible way;
- VSM allows a variety of similarity calculation methods, such as distance- or angle-based measures. This allows the comparison of query and a document, or a document and a document, revealing the properties of compared objects;
- Evaluation models such as the distance, angle, ellipse, conjunction, and disjunction are available to control a search in a vector space;
- The partial match technique can describe the degree of a matching between a query and a document representation. This can be used to rank retrieved documents on their correspondence to the query;
- Relevance feedback is essential for dynamically adjusting a search strategy. VSM allows this adjustment;
- VSM provides an environment for techniques like self-organiz-

ing maps, associative networks, multidimensional scaling, and distance- and angle-based visualization.

The weaknesses of VSM are:

- One problem is high dimensionality, which makes it hard to apply to a large document collection;
- Multiple features and attributes can be extracted from a document that can describe its subject. As the terms are extracted and used to construct a document-term matrix, the semantic contexts are lost;
- VSM assumes that all terms are independent. This may over-simplify the interrelationship between the term and its context.

Chapter Nine

Probabilistic Model of Information Retrieval

Introduction

Probabilistic Information Retrieval (PIR) computes the probability that a document is relevant to a query. Documents are ranked in decreasing order of relevance. Relevant documents are those whose prababilities of relevance in the ranked list exceed a cut-off value (Dominich 2000).

A formal definition of PIR is:

Let D be a set of objects (documents), $q \in D$ a query, $a \in \mathbb{R}$ a real cut-off value, and $P(R \mid (q, d))$ and $P(I \mid (q, d))$ the probability that document d is relevant (R) and irrelevant (I), respectively, to query q. It is assumed that $P(R \mid (d, d)) = 1$, $P(I \mid (d, d)) = 0$. The retrieved documents in response to query q belong to the set $\mathbf{R}(q)$ defined as follows:

$$\mathbf{R}(q) = \left\{ d \mid P\!\left(R \mid (q, d)\right) \geq P\!\left(I \mid (q, d)\right), \right.$$
$$\left. P\!\left(R \mid (q, d)\right) > a, \, a \in \mathbb{R} \right\}$$

The inequality $P\!\left(R \mid (q, d)\right) \geq P\!\left(I \mid (q, d)\right)$ is called Bayes' decision rule, and $P(R \mid (q, d))$ and $P(I \mid (q, d))$ are called the probability of relevance and irrelevance, respectively, of document d to query q. Users' information needs are translated into query representations. Documents are converted into document representations. IR systems try to determine whether documents meet information needs. IR systems have uncertain understanding of information needs. Probability theory provides a basis for computing the degree of matching and relevance (Manning, Raghavan, and Schütze 2008).

During searching and retrieval, there is a degree of guessing about relevance. To compute the accuracy of a guess, systems may use probability of relevance (Van Rijsbergen 1979), with a formula such as:

$$P_Q \ (relevance/document)$$

This computation of probability is based on frequency counts, which is a statistical and not a semantic approach. A matching function assigns a score to each document. Assuming one query has been submitted to the system, it will be represented as:

$$P \text{ (relevance/document)}$$

The relevance of a document is independent of other documents. The probability ranking principle states that a system's response to a request is the ranking of documents in order of decreasing probability of relevance.

The probability ranking principle assumes that we can calculate $P(relevance/document)$. Without knowing which are the relevant documents or how many there, there is no way to calculate $P(relevance/document)$. Using iteration, we can guess at $P(relevance/document)$. Assuming that each document is described by the presence or absence of terms, a document can be represented by a binary vector:

$$x = (x_1, x_2, \ldots, x_n)$$

where $x_i = 0$ or 1 indicates absence or presence of the i-th index term. There are two mutually exclusive potential events,

$$w_1 = relevant$$
$$w_2 = non\text{-}relevant$$

Probabilistic Model

There are both document-oriented and query-oriented retrieval models. VSM requires deciding whether to modify queries or to modify documents. Logical models must decide whether to estimate queries or documents or both. Regression and language models can estimate document or query characteristics, or both, and so on (Bodoff and Robertson 2004).

Early binary probabilistic approaches used relevance estimates for a particular document or query. Calculating a combined model begins with the assumption that the indexing is correct and has no error. That is, if a document D_i contains an arbitrary term t, and is indexed as $x_{it} = 1$, and if a query Q_i contains an arbitrary term t, $y_{it} = 1$, then term t correctly represents the meaning of the document and query. The probability of relevance is $x_{it} = 0/1$, $y_{jt} = 0/1$, e.g., $P(R = 1 \mid x_{it} = 0, y_{jt} = 1)$. Relevance is still uncertain for a particular document and query, even though they are correctly indexed.

To calculate the probability of relevance of D_k to Q_c, count relevance frequencies across all documents and queries. If all document indexes are correct, query indexes may have random error, in addition to the random error in relevancy data. Query indexes are random variables defined over the event space $\{Q\}$. The available data reflects two sources of random error.

Assuming that relevance, document indexing, and query indexing are all random variables with event spaces tries to account for all possible sources of error. Hidden variables of the correct indexing create the marginal views (query, relevance | correct documents) and (document, relevance | correct queries). There are many joint distributions of (document, query, relevance) that fit these two marginal distributions.

To use relevance data to estimate document and query parameters, an explicit error model is needed. Error models allow Bayesian updates or maximum likelihoods to adjust each parameter. More reliable data will rely on more adjustments from the relevance data, and less reliable data will adjust less (Bodoff and Robertson 2004).

The procedure requires three error models: document, query, and relevance.

Using the notation of a probability density as $f(w; \theta)$ with observations w and parameters θ, results in these three functions:

$$f_D(D_i^0; \underline{D}_i) = P(D_i^0 \mid \underline{D}_i),$$

$$f_Q(Q_i^0; \underline{Q}_i) = P(Q_i^0 \mid \underline{Q}_i), \text{ and}$$

$$f_R(R_{ij}; \underline{D}_i, \underline{Q}_j) = P(R_{ij} \mid \underline{D}_i, \underline{Q}_j)$$

This structure can be illustrated with the example of two documents and one query. $\underline{D}_i$ indicates the parameters of the probability function for one document. D_i^0 denotes indexing of the document. $\underline{D}_i$ generates D_i^0. Parameter $\underline{D}_i$ represents the meaning of the document in vector space. The function f_D represents the expression of an idea by an author. R_{ij} is the relevance value of a document-query pair, generated by the function whose parameters are a document vector and a query vector. Actual relevance depends on the actual meaning, which must be estimated. Relevance of a document-query pair is independent of the the observed words of texts (Bodoff and Robertson 2004).

Parameters and event spaces are not the same. The data-generating parameter for D_i^0 is $\underline{D}_i$, but the event space is the set of all documents, not the set of all $\underline{D}_i$. The difference is similar for queries. If two documents have the identical true parameter, it is not certain that their observations will be identical. Two authors can express the same idea with different words. The probability of relevance has been defined as $P(R_{ij} | \underline{D}_i, \underline{Q}_j)$, but there are two possibilities for the event space of a relevance judgment random variable, which are either the Cartesian product of all document and query parameters $\{\underline{D} \times \underline{Q}\}$, or the Cartesian product of all document and query individuals $\{D \times Q\}$. The latter implies that the parameter vectors do not capture all aspects of content, and that relevance depends on things besides "topicality." The conditional $P(R_{ij} | \underline{D}_i, \underline{Q}_j)$ means that the parameter values $\underline{D}_i, \underline{Q}_j$ limit the set of possible events to all pairs of document–query individuals with these parameter values. The event space of each relevance judgment may also be considered to be all pairs of document–query parameters. If so, then two identical pairs of document and query parameters will have identical relevance values. Document and query parameters are related in a particular but unknown way to an event that causes relevance (Bodoff and Robertson 2004).

Each observation and relevance value has a hypothesized distribution and a true parameter. The set of N_D documents D_i^0 is not a sample of size N_D from a single distribution, but a set of N_D random values, one D_i^0 from each of N_D distributions (Bodoff and Robertson 2004)

The functional form for each of $f_D(D_i^0; \underline{D}_i)$, $f_Q(Q_i^0; \underline{Q}_i)$, and $f_R(R_{ij}; \underline{D}_i, \underline{Q}_j)$ can be hypothesized. A likelihood function is constructed for all the data, in order to estimate the unknown parameter vectors $\underline{D}_i, \underline{Q}_j$ (Bodoff and Robertson 2004). After document and query parameters are estimated comes the prediction stage, where relevance predictions can use parameter vectors instead of the term vectors to predict the relevance of document-query pairs (Bodoff and Robertson 2004).

Using vectors that are real-valued with data consisting of three documents and two queries, suppose that of the six possible pairs, the following three are known to be relevant: (1, 1), (2, 2), (3, 2), while these three have unknown relevance: (2, 1), (3, 1), (1, 2). To estimate the true parameter vectors for all objects $\underline{D}_1, \underline{D}_2, \underline{D}_3, \underline{Q}_1, \underline{Q}_2$, it could be hypothesized that $P(D_1^0; \underline{D}_1)$, $P(D_2^0; \underline{D}_2)$, and $P(D_3^0; \underline{D}_3)$ are normal distributions with variance σ_D^2, that $P(Q_1^0; \underline{Q}_1)$ and $P(Q_2^0; \underline{Q}_2)$ have variance σ_Q^2, and that $P(R_{ij}; \underline{D}_i, \underline{Q}_j)$ is proportional to the inner products. That could result in the following hypothesized data: $\underline{D}_1, \underline{D}_2, \underline{D}_3, \underline{Q}_1, \underline{Q}_2$:

$$D_1{}^0 \sim N(\underline{D}_1, \sigma_D)$$
$$D_2{}^0 \sim N(\underline{D}_2, \sigma_D)$$
$$D_3{}^0 \sim N(\underline{D}_3, \sigma_D)$$
$$Q_1{}^0 \sim N(\underline{Q}_1, \sigma_Q)$$
$$Q_2{}^0 \sim N(\underline{Q}_2, \sigma_Q)$$
$$R_{11} \sim \underline{D}_1 \cdot \underline{Q}_1$$
$$R_{22} \sim \underline{D}_2 \cdot \underline{Q}_2$$
$$R_{32} \sim \underline{D}_3 \cdot \underline{Q}_2$$

If the data points are independent, a single integrated likelihood function can be constructed:

$$L = (\underline{D}_1 * \underline{Q}_1)(\underline{D}_2 * \underline{Q}_2)$$
$$\times (\underline{D}_3 * \underline{Q}_2)e^{-\frac{1}{2}\left(\frac{\underline{D}_1 - D_1{}^0}{\sigma_D}\right)^2} e^{-\frac{1}{2}\left(\frac{\underline{D}_2 - D_2{}^0}{\sigma_D}\right)^2} e^{-\frac{1}{2}\left(\frac{\underline{D}_3 - D_3{}^0}{\sigma_D}\right)^2}$$
$$\times e^{-\frac{1}{2}\left(\frac{\underline{Q}_1 - Q_1{}^0}{\sigma_Q}\right)^2} e^{-\frac{1}{2}\left(\frac{\underline{Q}_2 - Q_2{}^0}{\sigma_Q}\right)^2}$$

The estimated parameters $\hat{\underline{D}}_1, \hat{\underline{D}}_2, \hat{\underline{Q}}_1$ maximize the likelihood of the observations. After parameter estimation, predicting the relevance of the unknown pairs uses the same relevance function, but with estimated true parameters $R_{31} \sim \underline{D}_3 * \underline{Q}_1$ rather than the observed $D_3{}^0 * Q_1{}^0$. If a new document or query arrived, its observed $D_i{}^0$ ($Q_j{}^0$) would be matched against the estimated $\underline{Q}_j{}^0$ ($\underline{D}_i{}^0$) (Bodoff and Robertson 2004).

This combines document and query parameter estimation. Credit assignment is done based on principles and using probability. (Bodoff and Robertson 2004).

Part 3

Appendix

Chapter Ten

Mathematics in IR Instructional and Research Sectors

Fiji

University of the South Pacific
Faculty of Science and Technology
School of Computing, Information and Mathematical Sciences
http://www.usp.ac.fj/index.php?id=415

The School of Computing, Information, and Mathematical Sciences (SCIMS) is one of the largest schools in the University, and is well-resourced to pursue excellence in teaching and research. The Mission of SCIMS is to provide the best possible university education in mathematics, statistics, computing science and information systems to the people of the South Pacific. This includes the creation and transmission of knowledge in these areas. The School, via its four divisions, offers degrees in Computing Science, Information Systems, Mathematics and Statistics. It is rigorous in the implementation of current ICT-related technologies, with courses in data communication, computer networks and security, internet computing, data mining, project management, etc, being revised or developed on a timely basis. The School also pursues research vigorously, with many papers from staff published annually in leading international journals and conference proceedings.

Japan

Hiroshima University
Division of Mathematical and Information Sciences
http://www.mis.hiroshima-u.ac.jp/

This program aims at providing students with general understanding and the ability to solve problems in information and human behavior in contemporary society, by using methods of mathematical sciences, psychology, and political sciences. This Program consists of the Political

Sciences Curriculum, which offers ideas, systems, and methods of scientific and rational policy analyses; the Behavioral Sciences Curriculum, which provides scientific analyses of human behavior from a psychological viewpoint; and the Mathematical and Information Sciences Curriculum, which offers information sciences based on mathematical sciences and mathematical sciences with special emphasis of information sciences.

Japan

The University of Electro-Communications

Department of Computer Science and Information Mathematics
http://www.fedu.uec.ac.jp/JUSST/RD/Dept_CI.html

The Department offers Master's and doctoral degrees in computer science and engineering and in mathematical engineering related to information science. Faculty research subjects vary enormously from solid basic theories to modern challenging applications:
- **Computer Science:** Computer architecture, parallel and distributed processing, information theory, learning theory;
- **Software Engineering:** Algorithms, software systems, computer languages, natural language problems, color image processing;
- **Information Mathematics:** Basic mathematics, applied mathematics, numerical analysis, coding theory, mathematical problems in engineering; and
- **Applied Computer Science:** Artificial intelligence, linguistic information processing, image and speech processing, computer research in music and acoustics, computer aided instruction, mathematical programming.

Poland

Warsaw University of Technology
Faculty of Mathematics and Information Science
http://www.mini.pw.edu.pl/tiki-index.php?page=studies_en

Due to its strategic position in Europe, Poland has always been regarded as the gateway to the East. Warsaw is host to thousands of for-

eign companies doing business in Poland and many international institutions coordinating economic and scientific cooperation between, on the
one hand, the European Union and USA and on the other, Central and
Eastern European countries. This unique situation has created a high demand for well-qualified engineers fluent in English.

The Warsaw University of Technology is the highest-ranked institution for advanced engineering education and research in Poland and one
of the most prestigious academic institutions in Europe. The curricula
and academic standards closely resemble those of many highly regarded
U.S. universities. The low ratio of students to professors at Warsaw University of Technology's Faculty of Mathematics and Information Science
and the warm working relationship between students and instructors
helps those studying to develop confidence in their ability to make significant contributions to engineering and research work.

The Faculty of Mathematics and Information Science offers the courses
in Mathematics and Computer Science in Polish. The following programs
are taught in English however:

- 3.5-year undergraduate program in Computer Science leading to
  a Bachelor of Science in Engineering degree; and
- 2-year graduate program in Computer Science leading to a Master of Science degree.

Starting from the 2008/2009 academic year, the 1.5-year graduate program in Computer Science leading to a Master of Science degree will commence. Studies will start every half year - in October and in February.

Graduate programs exist in the following fields:

- Artificial Intelligence;
- Computing in Business and Economics; and
- Computing in Science and Engineering.

In each of the above three specializations there are general lectures on
modern databases, foundations of Artificial Intelligence, Windows programming, operating systems, computer network administration as well
as several courses particular to the area of specialization. Each year, depending on the number of candidates, one or two programs commence,
depending on students' preferences.

Students graduating in studies taught in the English language will receive their diplomas in the Polish language and a translation into English
where:

- The Polish professional degree of an engineer is translated as a Bachelor of Science in Engineering; and
- The Polish professional degree of a master is translated as a Master of Science.

The programs taught in English follow an English/American pattern and lead to a Bachelor of Science in Engineering degree in 3.5 years and a Master of Science degree in a further two years. The duration and extent of the programs are defined by a system of credits. Credits are obtained on the basis of teaching results and are attained by a grade which is based on the semester workload and/or oral/written examinations. A student is supposed to study 30 credit points per semester (60 credits per academic year). The Bachelor's degree programs total 210 credits over seven semesters. To proceed from one year to the next, students have to obtain a certain minimum amount of credits defined by the Faculty Council. On successful completion of the final semester, final thesis and on passing the diploma examination students will be granted a B.Sc. degree. The Master's degree programs total 120 credits over four semesters (two years) or 90 credits over 3 semesters.

Russia

R.Y. Alekseev Nizhny Novgorod State Technical University
Institute of Radio Engineering and Information Technologies (IRIT)
http://www.nntu.ru/NSTU/facul/facul_spec/irit.htm

The training and research Institute of Radio Engineering and Information Technologies (IRIT) has been established on the order No. 200 from December 21st, 2005 on the basis of the Faculty of the Information Systems and Technologies.

In the past 70 years the Faculty and later the Institute have accumulated wide experience of training engineers and research personnel particularly appreciated both in Russia and abroad.

Seven Lenin prize winners, more than 50 Laureates of the State Prize, dozens of Doctors of Engineering and hundreds of Candidates of Science (Engineering) as well as research and engineering managerial staff of the Nizhny Novgorod's largest branch research institutes and telecommunication companies.

The Institute of Radio Engineering and Information Technologies provides training in the following degree courses and professions:

- Radio Engineering;
- Design and technology of electronic facilities (electronic instrumentation);
- Information Science and Computer Engineering;
- Computer-aided Data-processing and Control Systems;
- Information Management Systems and Technologies;
- Information-processing Technologies for Education;
- Information-processing Technologies for Design;
- Applied Mathematics;
- Telecommunications;
- Networking and Switching Systems; and
- Wireless communication, broadcasting and television.

Slovenia

University of Ljubljana
Faculty of Computer and Information Science
Laboratory for Mathematical Methods in Computer and Information Science
http://www.fri.uni-lj.si/en/laboratories/mathematics_physics_group/lab_mathematical_methods_computer_information_science/

The research activities of the laboratory involve various fields of mathematics with special emphasis on applications to computer and information science. The following areas of mathematics are studied: scientific computing and numerical solutions of differential equations, in particular, methods for geometric integration, graph theory, mostly topological and structural properties of graphs, vertex colorings of graphs and weighted graphs as a natural generalization of the channel assignment problem, algebraic topology, in particular cohomology of topological spaces with group actions, applications of topology to computer science, and computational topology, nonlinear dynamical systems and their application in geometry, physics and mechanics, linear and nonlinear mathematical techniques in computer vision (in cooperation with the Computer vision laboratory), computational geometry and geometry of cycles (in cooperation the Faculty of Electrical Engineering and the Faculty of Mathematics and Physics) with applications to surface modeling, in the area of incidence structures we study problems related to combinatorial and geometric configurations (the study of combinatorial properties of configurations via their incidence graphs, and the study of possibility of the realization of configurations in other incidence structures), CFD programs and their use in sailing simulations

The laboratory organizes the Mathematical seminar at the FRI, where members of the lab and other researchers report on current work, connected to the research and teaching activities of the lab. Several members of the lab are also members of research groups of the Institute of Mathematics, Physics, and Mechanics.

Members of the lab are involved in joint research work with other research groups at the Faculty of Computer and Information Science and the Faculty of Electrical Engineering and with the following institutions: NTNU Trondheim, Norway, and University in Bergen, Norway.

UK

University of Brighton
Computing, Mathematical and Information Sciences
http://www.brighton.ac.uk/cmis/contact/details.php?uid=je11

The School of Computing, Mathematical and Information Sciences is at the forefront of teaching, research and consultancy in a range of discipline areas:

- Computing;
- Information and library studies;
- Mathematical sciences; and
- Media and communication.

The school offers a comprehensive range of foundation degree, honors degree, PhD and masters degree programs. Extensive applied research and consultancy activity ensures that these courses are up to date, relevant and serve the needs of both students and employers.

The school is located in the Watts building of the Moulsecoomb site, about 2 miles from the sea front, pier, and Brighton city center, with library, study and computing facilities catered for within the campus.

USA

Jacksonville State University
College of Arts and Sciences
Department of Mathematical, Computing, and Information Sciences
http://mcis.jsu.edu/

As part of the College of Arts and Sciences at Jacksonville State University, the MCIS Department offers undergraduate degrees (B.S.) in Mathematics, Computer Science, and Computer Information Systems. Master's degrees (M.S.) in Mathematics as well as in Computer Systems and Software Design are also offered.

The Department of Mathematical, Computing, and Information Sciences is accredited by the Southern Association of Colleges and Schools (SACS) with the Bachelor of Science degrees in Computer Science and Computer Information Systems additionally accredited by the Computing Accreditation Commission (CAC) of the Accreditation Board for Engineering and Technology (ABET).

The mission of the Mathematical, Computing, and Information Sciences Department is to provide quality programs for three Bachelor of Science degrees and two Master of Science degrees. The MCIS Department also provides a minor in four programs and provides the courses for students seeking a Bachelor of Science in Education with a concentration in secondary mathematics. The programs of this department place a strong emphasis on quality teaching and encourage research and other scholarly activities to strengthen this emphasis. In its presentation of the programs, the department attempts to provide a balance between the theory and the practice.

The MCIS Department provides the appropriate courses for its own majors and minors in addition to several service courses in Mathematics and Computer Science for other disciplines. In conjunction with Academic Computer Services, the department attempts to provide all its students access to modern computing laboratories.

- The Major Field Test (MFT) for Computer Science and for Mathematics is administered;
- An Exit Survey and Senior Exit Interviews to graduating seniors in each of our major areas (CS and MS) in the fall and spring are administered;

- An alumnus questionnaire specific to our fields, which accompanies the University alumni questionnaire, is administered; and
- The students are tracked in the MS 100 and MS 112 math classes.

USA

Mercy College
Mathematics and Computer Information Science
http://www.mercy.edu/acadivisions/mathcompinfo/index.cfm

The degree from the Mathematics Program at Mercy College places its emphasis on applied mathematics. The goal of the program is to prepare the students for success in any field that requires mathematics, e.g., actuaries, banking, and finance. The student will also be prepared for graduate programs in several fields including mathematics, computer sciences, environmental sciences, and biology. Students in this major will learn at least one computer language. They will receive a strong foundation in applied mathematics through Statistics and probability, Linear Algebra, Mathematical Modeling, Differential Equations, Numerical Analysis, and Algebraic Structures. Computers and graphing calculators are used extensively in most of the math courses.

The Computer Information Science program at Mercy College is designed to enable students to adapt to fast-changing technologies by:

- Introducing students to modern technology and software;
- Developing competencies, skills and a knowledge base for graduates that are necessary for success in the work place and lifelong learning; and
- Preparing majors for graduate programs in related fields including Computer Science, Computer Information Systems, Telecommunication, Management Information Systems, Information Technology, Internet Business Systems, and Business.

USA

University of Illinois at Chicago
Department of Mathematics, Statistics, and Computer Science
Mathematical and Information Sciences for Industry (MISI)
http://www.math.uic.edu/~misi/approach.html

UIC's Master's degree program in Mathematical and Information Sciences for Industry (MISI) is unique in providing:

- An interdisciplinary curriculum balancing discrete mathematics, continuous mathematics, computer science, information sciences, and statistics. To successfully model complex scientific, engineering and business problems, students need a knowledge not only of discrete mathematics, including algorithms, combinatorics, data structures, and programming, but also of classical topics in applied mathematics, including differential equations, numerical analysis, mathematical modeling, and analysis. This curriculum balances discrete mathematics, which forms the foundation of a computer science curriculum, with the continuous mathematics used to model complex scientific, engineering, and business problems. In addition, students receive a solid foundation in software science, computer modeling, and simulation;
- The opportunity for students to work in teams on advanced research and development projects. Each student will complete a major project in his or her area of specialization and several minor projects. Projects allow students to work in depth on a problem of interest to both themselves and industry. Projects also allow students to see current, ongoing research in new and emerging fields, as well as those fields containing a strong computational component; and
- Workshops and projects to improve student's technical oral and written communications skills. Workshops and projects provide students with the opportunity to learn to work in a team and to improve their technical oral and written communication skills.

The program's approach is to have students concentrate on a core set of required Industrial Mathematics concentration courses and on a curriculum which is project-oriented. As outlined below, concentration courses are required in the areas of Discrete Mathematics, Applied Mathematics, Information and Software Sciences, Statistics and Probability, and Industrial Science. In addition, students graduating from the program will have worked as team members on one major project and several minor projects; each project chosen based on its applicability to current industrial needs and based on its ability to provide results which are meaningful to industry.

The program has been designed to focus simultaneously on the student graduating from the program and the industry that will potentially hire him or her. One focus is on the graduate student seeking a master's degree in mathematics but intending to seek employment in a non-academic career. The other is on companies in need of employees with an advanced degree in industrial mathematics and/or in need of UIC's expertise to assist them in finding solutions to projects of many types.

Graduate Students

The one and a half year MISI Master's Degree Program is structured so that graduates will have developed three overlapping skill sets:

- The breadth and depth of their *mathematical knowledge* will allow them to make meaningful contributions to the solutions of complex problems requiring sophisticated analysis;
- Their knowledge of *computer science* will give them the ability to develop algorithms and software which allow "real world" solutions to complex problems; and
- Their oral and written *communication skills,* along with project management skills, will allow them to formulate and express easily understood technical goals and to insert new technology into an organization.

For these reasons, the program is interdisciplinary and places equal emphasis on mathematics, information sciences, oral and written communication skills, and project management.

- Business   and   Industrial   Participation UIC envisions three ways in which companies can participate in the MISI program:
- Companies can sponsor one of their own current employees who enrolls in the program;
- Companies can submit projects to one of the department's laboratories; and
- Companies can sponsor ongoing research undertaken by MISI program graduate research assistants in one of the department's laboratories.

USA

US Army Research Laboratory
ARO Mathematical and Information Sciences
http://www.arl.army.mil/www/default.cfm?Action=29&Page=216

The mission of the US Army Research Laboratory is to support and sponsor basic research in the information processing, computation, and mathematical modeling, enhance decision-making, command and control, communications, and combat system performance.

References

Works Cited

Bodoff, D., & Robertson, S. (2004). A new unified probabilistic model. *Journal of the American Society for Information Science & Technology* 55 (6): 471-487.

Dominich, S. (2000). A unified mathematical definition of classical information retrieval. *Journal of the American Society for Information Science* 51 (7): 614-624.

Dominich, S. (2001). *Mathematical foundations of information retrieval.* Dordrecht: Kluwer.

Góth, J., & Skrop, A. (2005). Varying retrieval categoricity using hyperbolic geometry. *Information Retrieval* 8 (2): 265-283.

Kowalski, G. (1997). *Information retrieval systems: Theory and implementation.* Norwell, MA: Kluwer.

Manning, C. D., Raghavan, P., & Schütze, H. (2008). *Introduction to information retrieval.* Cambridge: Cambridge University Press.

Park, L. A. F., Ramamohanarao, K., & Palaniswami, M. (2005). A novel document retrieval method using the discrete wavelet transform. *ACM Transactions on Information Systems* 23 (3): 267-298.

van Rijsbergen, C. J. (1979). *Information retrieval.* Glasgow: University of Glasgow.

Zhang, J. (2008). *Visualization for information retrieval.* Milwaukee: Springer.

Works Consulted

Alimohammadi, D. (2006). The equilateral triangle paradigm: A mathematical interpretation of the theory of tertiary sources on the World Wide Web. *Library Philosophy and Practice* 8(2): Available at: http://unllib.unl.edu/LPP/alimohammadi.htm

Angelova, R., & Weikum, G. (2006). Graph-based text classification: learn from your neighbors. *SIGIR Forum* 39: 485-492.

Anh, V. N., & Moffat, A. (2006). Improved word-aligned binary compression for text indexing. *IEEE Transactions on Knowledge & Data Engineering* 18 (6): 857-861.

Aslam, J. A., Yilmaz, E., & Pavlu, V. (2005). The maximum entropy method for analyzing retrieval measures. *SIGIR Forum*. 27-34.

Baeza-Yates, R., & Ribeiro-Neto, B. (1999). *Modern information retrieval*. New York: Addison-Wesley.

Barbay, J., Golynski, A., Munro, J. I., & Rao, S. S. (2007). Adaptive searching in succinctly encoded binary relations and tree-structured documents. *Theoretical Computer Science* 387 (3): 284-297.

Bell, D., McErlean, F., Stewart, P., & McClean, S. (1990) Application of Simulated Annealing to Clustering Tuples in Databases. *Journal of the American Society for Information Science* 41 (2): 98-110.

Berry, M. W., Browne, M., Langville, A. N., Pauca, V. P., & Plemmons, R. J. (2007). Algorithms and applications for approximate nonnegative matrix factorization. *Computational Statistics & Data Analysis* 52 (1): 155-173.

Beynon, M., Curry, B., & Morgan, P. (2000). Classification and rule induction using rough set theory. *Expert Systems* 17 (3): 136-148.

Blanchard, A. (2007). Understanding and customizing stopword lists for enhanced patent mapping. *World Patent Information* 29 (4): 308-316.

Broder, A. Z., Lempel, R., Maghoul, F., & Pedersen, J. (2006). Efficient pagerank approximation via graph aggregation. *Information Retrieval* 9 (2): 123-128.

Cai, D., He, X., & Han, J. (2005). Document clustering using locality preserving indexing. *IEEE Transactions on Knowledge & Data Engineering* 17 (12): 1624-1637.

Canny, J. (2004). GaP: A factor model for discrete data. *SIGIR Forum*. 122-129.

Chang, Y., Kim, M., & Raghavan, V. V. (2006). Construction of query concepts based on feature clustering of documents. *Information Retrieval* 9 (3): 231-248.

Cheng, C. P., Lau, G. T., Law, K. H., Pan, J., & Jones, A. (2009). Improving access to and understanding of regulations through taxonomies. *Government Information Quarterly* 26 (2): 238-245.

Chomicki, J. (2007). Semantic optimization techniques for preference queries. *Information Systems* 32 (5): 670-684.

Diaz, F., & Metzler, D. (2006). Improving the estimation of relevance models using large external corpora. *SIGIR Forum* 39: 154-161.

Diligenti, M., Gori, M., & Maggini, M. (2004). A unified probabilistic framework for web page scoring systems. *IEEE Transactions on Knowledge & Data Engineering* 16 (1): 4-16.

Falchi, F., Gennaro, C., & Zezula, P. (2008). Nearest neighbor search in metric spaces through content-addressable networks. *Information Processing & Management* 44 (1): 411-429.

Ferguson, I. A., & Durfee, E. H. (1998). Artificial intelligence in digital libraries: Moving from chaos to (more) order. *International Journal on Digital Libraries* 2 (1): 309.

Fide, R. (1991). Searchers' selection of search keys: III. Searching styles. *Journal of the American Society for Information Science* 42 (7): 515-527.

Gordon, M., & Kochen, M. (1989). Recall-precision trade-off: A derivation. *Journal of the American Society for Information Science* 40 (3): 145-151.

Harmon, G. (2008). Remembering William Goffman: Mathematical information science pioneer. *Information Processing & Management* 44 (4): 1634-1647.

Jin, R., & Si, L. (2004). A study of methods for normalizing user ratings in collaborative filtering. *SIGIR Forum.* 568-569.

Kantor, P. B. (2002). Mathematical models in information science. *Bulletin of the American Society for Information Science & Technology* 28 (6):

Kowalski, G. J., & Maybury, M. T. (2002). *Information storage and retrieval systems: Theory and implementation.* New York: Kluwer Academic Publishers.

Kuflik, T., Shapira, B., & Shoval, P. (2003). Stereotype-based versus personal-based filtering rules in information filtering systems. *Journal of the American Society for Information Science & Technology* 54 (3): 243-250.

Larson, R. (1992). Letters to the Editor. *Journal of the American Society for Information Science* 43 (7): 518.

Lee, M., & Corlett, E. (2003). Sequential sampling models of human text classification. *Cognitive Science* 27 (2): 159.

Lee, M., & Pun, C. (2003). Rotation and scale invariant wavelet feature for content-based texture image retrieval. *Journal of the American Society for Information Science & Technology* 54 (1): 68.

Li, S., Lee, M., & Adjeroh, D. (2005). Effective invariant features for shape-based image retrieval. *Journal of the American Society for Information Science & Technology* 56 (7): 729-740.

Li, T., Zhu, S., & Ogihara, M. (2008). Text categorization via generalized discriminant analysis. *Information Processing & Management* 44 (5): 1684-1697.

Lin, J., Wu, P., Demner-Fushman, D., & Abels, E. (2006). Exploring the limits of single-iteration clarification dialogs. *SIGIR Forum* 39: 469-476.

Lin, M.Y., & Lee, S. Y. (2004). Incremental update on sequential patterns in large databases by implicit merging and efficient counting. *Information Systems* 29 (5): 385-404.

Lin, R. S., & Chen, L. H. (2005). Content-based audio retrieval based on gabor wavelet filtering. *International Journal of Pattern Recognition & Artificial Intelligence* 19 (6): 823-837.

Lin, X. (1997). Map displays for information retrieval. *Journal of the American Society for Information Science* 48 (1): 40-54.

Losada, D., & Barreiro, A. (2003). Embedding term similarity and inverse document frequency into a logical model of information retrieval. *Journal of the*

American Society for Information Science & Technology 54 (4): 285.

Luk, R. (2008). On event space and rank equivalence between probabilistic retrieval models. *Information Retrieval*. 11 (6): 539-561.

MacPherson, D. L. (2006). Digitizing the non-digital: Creating a global context for events, artifacts, ideas, and information. *Information Technology & Libraries* 25 (2): 95-102.

McCarn, D. B., & Lewis, C. M. (1990). A mathematical model of retrieval system performance. *Journal of the American Society for Information Science* 41 (7): 495-500.

Meo, R. (2000). Theory of dependence values. *ACM Transactions on Database Systems* 25 (3): 380-406.

Min, Z., GuoDong, Z., & Aiti, A. (2008). Exploring syntactic structured features over parse trees for relation extraction using kernel methods. *Information Processing & Management* 44 (2): 687-701.

Montesi, D., Trombetta, A., & Dearnley, P. (2003). A similarity based relational algebra for Web and multimedia data. *Information Processing & Management* 39 (2): 307.

Norberg, L. R., Vassiliadis, K., Ferguson, J., & Smith, N. (2005). Sustainable design for multiple audiences: The usability study and iterative redesign of the Documenting the American South digital library. *OCLC Systems & Services* 21 (4): 285-299.

Nuntiyagul, A., Naruedomkul, K., Cercone, N., & Wongsawang, D. (2007). Keyword extraction strategy for item banks text categorization. *Computational Intelligence* 23 (1): 28-44.

Oh, S. (1998). Document representation and retrieval using empirical facts: Evaluation of a pilot system. *Journal of the American Society for Information Science* 49 (10): 920-931.

Olsson, J. S. (2006). An analysis of the coupling between training set and neighborhood sizes for the kNN classifier. *SIGIR Forum* 39: 685-686.

Pavlov, D., Manavoglu, E., Giles, C. L., & Pennock, D. M. (2004). Collaborative filtering with maximum entropy. *IEEE Intelligent Systems* 19 (6): 40-48.

Perlovsky, L. (2007). Cognitive high level information fusion. *Information Sciences* 177 (10): 2099-2118.

Pestov, V., & Stojmirović, A. (2006). Indexing schemes for similarity search: An illustrated paradigm. *Fundamenta Informaticae* 70 (4): 367-385.

Plachouras, V., & Ounis, I. (2005). Dempster-Shafer Theory for a query-biased combination of evidence on the Web. *Information Retrieval* 8 (2): 197-218.

Rein, S., & Reisslein, M. (2006). Identifying the classical music composition of an unknown performance with wavelet dispersion vector and neural nets. *Infor-*

mation Sciences 176 (12): 1629-1655.

Sindhwani, V., & Keerthi, S. S. (2006). Large scale semi-supervised linear SVMs. *SIGIR Forum* 39: 477-484.

Smucker, M. D., & Allan, J. (2006). Find-similar: Similarity browsing as a search tool. *SIGIR Forum* 39: 461-468.

Sokolova, M. (2009). A systematic analysis of performance measures for classification tasks. *Information Processing & Management* 45 (4): 427-437.

Subrt, T., & Brozova, H. (2007). Knowledge maps and mathematical modelling. *Electronic Journal of Knowledge Management* 5 (4): 497-503.

Tan, C. L., & Cao, R. (2002). Restoring of archival documents using a wavelet technique. *IEEE Transactions on Pattern Analysis & Machine Intelligence* 24 (10): 1399.

Taniguchi, S. (2004). Design of cataloging rules using conceptual modeling of cataloging process. *Journal of the American Society for Information Science & Technology* 55 (6): 498-512.

Tao, T., & Zhai, C. X. (2006). Regularized estimation of mixture models for robust pseudo-relevance feedback. *SIGIR Forum* 39: 162-169.

Vakkari, P., & Sormunen, E. (2004). The influence of relevance levels on the effectiveness of interactive information retrieval. *Journal of the American Society for Information Science & Technology* 55 (11): 963-969.

Valle-Lisboa, J. C., & Mizraji, E. (2007). The uncovering of hidden structures by Latent Semantic Analysis. *Information Science* 177 (19): 4122-4147.

van Rijsbergen, C. J. (2004). *The geometry of information retrieval.* Cambridge: Cambridge University Press.

Vinay, V., Cox, I. J., Milic-Frying, N., & Wood, K. (2006). On ranking the effectiveness of searches. *SIGIR Forum* 39: 398-404.

Wang, J. Z., Wiederhold, G., Firschein, O., & Wei, S. X. (1998). Content-based image indexing and searching using Daubechies' wavelets. *International Journal on Digital Libraries* 2 (1): 311-328.

Warren, S. (2001). Visual displays of information: A conceptual taxonomy. *Libri: International Journal of Libraries & Information Services.* 51 (3): 135-147.

Williams, P. (2005). Where do I start? A cartographic cataloguing code. *Cartographic Journal* 42 (3): 227-230.

Zhang, J., Wolfram, D. (2001). Visualization of term discrimination analysis. *Journal of the American Society for Information Science & Technology* 52 (8): 615-627.

Zhang, Y. (2005). Bayesian graphical models for adaptive filtering. *SIGIR Forum* 39 (2): 57.

Zhu, S., Ji, X., Xu, W., & Gong, Y. (2005). Multi-labelled classification using maximum entropy method. *SIGIR Forum* 274-281.

Further Reading

Baxendale, P., & Korfhage, R. (1968). A note on a relevance estimate and its improvement. *Communications of the ACM* 11 (11): 756.

Bodoff, D. (2004). Relevance models to help estimate document and query parameters. *ACM Transactions on Information Systems* 22 (3): 357-380.

Brandhorst, W. (1970). A table of set theory notations. *Journal of the American Society for Information Science* 21 (6): 428.

Burkowski, F. (1992). An algebra for hierarchically organized text-dominated databases. *Information Processing & Management* 28 (3): 333-348.

Chazelle, B. (1990). Lower bounds for orthogonal range searching. II. The arithmetic model. *Journal of the Association for Computing Machinery* 37 (3): 439-463.

Flores, I., & Madpis, G. (1971). Average binary search length for dense ordered lists. *Communications of the ACM* 14 (9): 602.

Harabagiu, S., Lacatusu, F., & Hickl, A. (2006). Answering complex questions with random walk models. *SIGIR Forum* 39: 220-227.

Harter, S. (1971). Optimum number of cards per file guide assuming binary searching. *Journal of the American Society for Information Science* 22 (2): 140.

Kaplan, F., & Hafner, V. (2006). Information-theoretic framework for unsupervised activity classification. *Advanced Robotics* 20 (10): 1087-1103.

Kokare, M., Biswas, P., & Chatterji, B. (2005). Texture image retrieval using new rotated complex wavelet filters. *IEEE Transactions on Systems, Man & Cybernetics: Part B* 35 (6): 1168-1178.

Kolda, T., & O'Leary, D. (1998). A semidiscrete matrix decomposition for latent semantic indexing in information retrieval. *ACM Transactions on Information Systems* 16 (4): 322.

Lovins, J. (1971). Error evaluation for stemming algorithms as clustering algorithms. *Journal of the American Society for Information Science* 22 (1): 28-40.

Lowe, T. (1971). Effectiveness of retrieval key abbreviation schemes. *Journal of the American Society for Information Science* 22 (6): 374-381.

Maarek, Y., Berry, D., & Kaiser, G. (1991). An information retrieval approach for automatically constructing software libraries. *IEEE Transactions on Software Engineering* 17 (8): 800-813.

March, S. (1983). Techniques for structuring database records. *ACM Computing Surveys* 15 (1): 45.

Mazur, Z. (1988). Properties of a model of distributed homogeneous information retrieval system based on weighted models. *Information Processing & Management* 24 (5): 525-540.

Munavalli, R., & Miner, R. (2006). MathFind: A math-aware search engine. *SIGIR Forum* 39: 735.

Nance, R., Korfhage, R., & Bhat, U. (1972). Information networks: Definitions and message transfer models. *Journal of the American Society for Information Science* 23 (4): 237-247.

Ord, J., & Iglarsh, H. (2007). The estimation of conditional distributions from large databases. *American Statistician* 61 (4): 308-314.

Pasterczyk, C. (1990). Mathematical notation in bibliographic databases. *Database Magazine* 13 (4): 45-56.

Pirrò, G., & Talia, D. (2008). LOM: A linguistic ontology matcher based on information retrieval. *Journal of Information Science* 34 (6): 845-860.

Pogorelec, A., & Šauperl, A. (2006). The alternative model of classification of belles-lettres in libraries. *Knowledge Organization* 33 (4): 204-214.

Robertson, S., & Maron, M. (1983). The unified probabilistic model for IR. *Lecture Notes in Computer Science* 146.

Roelleke, T., & Wang, J. (2006). A parallel derivation of probabilistic information retrieval models. *SIGIR Forum* 39: 107-114.

Rush, J., & Russo, P. (1971). A note on the mathematical basis of the SLIC index. *Journal of the American Society for Information Science* 22 (2): 123-125.

Sagiv, Y. (1988). On bounded database schemes and bounded Horn-clause programs. *SIAM Journal on Computing* 17 (1): 1-22.

Schonfeldt, R. (1994). Mathematical characteristics of thesaurus relations (English). *Nachrichten fur Dokumentation* 45 (4): 203-212.

Shishibori, M., & Koyama, M. (2000). Two improved access methods on compact binary (CB) trees. *Information Processing & Management* 36 (3): 379.

Swanson, R. (1972). On terminology for search strategies. *Journal of the American Society for Information Science* 23 (2): 133-134.

Tan, K. (1972). On Foster's information storage and retrieval using AVL Trees. *Communications of the ACM* 15 (9): 843.

van Rijsbergen, C. J. (2006). Quantum haystacks. *SIGIR Forum* 39: 1-2.

Wang, X., Tao, T., & Sun, J. (2008). Rank: Solving the zero-one gap problem of page rank. *ACM Transactions on Information Systems* 26 (2): 10:1-10:29.

Index

Nebraska
UNIVERSITY OF
Lincoln